HUMAN AUGMENTATION

A GUIDE TO THE FUTURE OF OUR BODIES AND MINDS.

A NEW ERA OF HUMAN EVOLUTION.

CONTENTS

CHAPTER 1: INTRODUCTION

Defining Human Augmentation:

- What does it mean to augment our bodies and minds? We will explore the various definitions and differentiate between different types of enhancements, such as:

Prosthetic devices: Replacing lost limbs or senses with artificial counterparts

Implantable technologies: Integrating electronic devices within the body to enhance physical or cognitive abilities

Biotechnological modifications: Gene editing, stem cell therapies, and other interventions impacting biological functions

Cognitive enhancements: Utilizing drugs, brain stimulation, or technology to improve memory, focus, or learning

A Brief History of Enhancement Technologies:

- We will take a historical journey to understand how humans have always sought ways to enhance themselves, from ancient practices like trephination to the development of eyeglasses and early prosthetics.
- The chapter will explore pivotal moments in the history of augmentation, highlighting significant advancements and their impact on society.

The Ethical Landscape:

- This crucial section dives into the complex ethical questions surrounding human augmentation.

We will discuss various viewpoints and considerations, including:

Fairness and equality: Will enhancements exacerbate existing inequalities and create an "augmented elite"? How can we ensure equitable access to these technologies?

Informed consent and autonomy: Should individuals have the right to choose augmentation, even if it carries risks? How can we ensure informed decisions considering the potential for pressure or coercion?

The definition of human: Does augmentation blur the lines between human and machine? How will it impact our self-perception and societal values?

Safety and unintended consequences: What are the potential risks associated with these technologies, both physical and psychological? How can we ensure responsible development and mitigate unforeseen consequences?

Military applications and societal control: How could human augmentation be used in warfare or for social control? How can we safeguard against its misuse?

CHAPTER 2: EXPLORING THE SPECTRUM OF ENHANCEMENTS

1. Physical Augmentation: From Prosthetics to Superhuman Strength

- **Prosthetics:** We will explore the advancements in artificial limbs and body parts, their functionalities, and the impact they have on individuals with disabilities, improving mobility, dexterity, and overall well-being.

- **Exoskeletons and Wearable Augmentation:** This section will discuss wearable technologies that enhance physical capabilities, such as strength, endurance, and speed. We will explore their diverse applications in fields like rehabilitation, construction, and even military use.

- **Neural Implants and Brain-Computer Interfaces (BCIs):** We will delve into how BCIs directly connect the brain with external devices, bypassing the limitations of the body. This section will explore various applications, including controlling prosthetic limbs, restoring lost senses, and potentially even enhancing cognitive abilities.
- **Implantable Technologies:** This section discusses various implants that enhance or replace bodily functions, such as pacemakers, cochlear implants, and potential future advancements like artificial organs. We will explore the ethical considerations and potential risks associated with these technologies.

- **Pushing the Boundaries: Superhuman Strength and Beyond:** While currently hypothetical, this section explores the possibilities of augmentations that fundamentally alter human physical capabilities, pushing the boundaries of strength, speed, and endurance. We will discuss the ethical considerations and potential implications of such advancements.

2. Cognitive Enhancement: Boosting Memory, Focus, and Learning

- **Nootropics and Cognitive Drugs:** This section explores medications and supplements aimed at improving memory, focus, and cognitive function. We will discuss their efficacy, potential side effects, and the ethical considerations surrounding their use, especially in healthy individuals.

- **Brain Stimulation Techniques:** We will delve into various methods of brain stimulation, such as transcranial magnetic stimulation (TMS) and deep brain stimulation (DBS), and their potential applications in enhancing cognitive function, treating neurological disorders, and even improving mental health.
- **Brain-Computer Interfaces and Cognitive Augmentation:** Building upon the discussion from the physical augmentation section, this section explores how BCIs can potentially enhance cognitive abilities directly interacting with the brain. We will discuss the potential applications in areas like memory augmentation, learning

enhancement, and even treatment of mental illnesses.

- **Ethical Considerations:** This section emphasizes the importance of responsible development and ethical considerations surrounding cognitive enhancements. We will address concerns about dependence, cognitive over-reliance, and the potential for widening the gap between individuals with and without access to such enhancements.

3. Sensory Augmentation: Expanding the Boundaries of Perception

- **Prosthetics for Sensory Loss:** This section discusses how advancements in prosthetics can restore lost senses, such as cochlear implants for hearing and retinal implants for vision. We will delve into the challenges and potential benefits of these technologies, highlighting the impact they have on individuals' lives.

- **Sensory Augmentation for Healthy Individuals:** We will explore technologies that enhance existing senses beyond their natural limitations, such as night vision goggles or devices that amplify hearing. We will discuss the potential benefits and risks of pushing the boundaries of human perception.

- **Brain-Computer Interfaces and Sensory Enhancement:** Building upon previous chapters, this section explores how BCIs can potentially bypass traditional sensory organs and directly stimulate the brain to create new sensory experiences or enhance existing ones. We will discuss the potential applications and ethical considerations surrounding this emerging field.

4. Emotional and Social Augmentation:

- **Brain Stimulation for Mental Health:** This section explores how brain stimulation techniques like DBS and TMS are being explored for treating various mental health conditions, such as depression and anxiety. We will discuss the potential benefits and limitations of these approaches.
- **Neurofeedback and Emotional Regulation:** We will delve into neurofeedback, a technique that provides individuals with real-time feedback on their brain activity, allowing them to learn how to

regulate their emotions. We will explore its potential applications for emotional well-being and mental health.

- **Social Augmentation and Human-Machine Communication:** This section explores future possibilities of technologies that enhance our ability to communicate, understand, and connect with others, potentially even bridging the gap between humans and machines. We will discuss the ethical considerations surrounding such advancements and the potential impact on social interactions.

CHAPTER 3: THE SCIENCE BEHIND THE ENHANCEMENTS

1. Bioengineering and Genetic Editing:

- **Biomaterials:** We begin by exploring the world of biomaterials, synthetic materials designed to interact with living tissues in a compatible and functional manner. This section will discuss the development of advanced materials for prosthetics, implants, and potentially even organ regeneration. Biocompatible materials are crucial for ensuring the long-term success of implants and avoiding rejection by the body. Scientists are constantly innovating in this field, creating materials with improved strength, flexibility, and biocompatibility, pushing the boundaries of what is possible in human augmentation.

- **Tissue Engineering and Stem Cell Therapy:** We will explore how scientists are utilizing stem cells, the body's master cells with the potential to develop into various specialized cell types, to grow and repair tissues. This holds immense promise for treating various diseases and injuries, potentially offering solutions for conditions like heart disease, diabetes, and spinal cord injuries. The ethical considerations surrounding stem cell research, particularly regarding the source of stem cells and the potential for misuse, will also be addressed in this section.

- **Gene Editing Technologies:** This section delves into revolutionary tools like CRISPR-

Cas9, which allows scientists to edit genes with unprecedented accuracy. We will discuss the potential applications of gene editing in various aspects of human augmentation:

Treating genetic diseases: Gene editing offers the possibility of correcting faulty genes responsible for numerous inherited diseases, potentially leading to cures for conditions like cystic fibrosis, sickle cell anemia, and Tay-Sachs disease.

Enhancing physical and cognitive abilities: While the ethical implications are significant, some researchers are exploring the potential of gene editing to enhance aspects like muscle strength, endurance, or even cognitive function. However, the potential risks and unintended consequences of such interventions need careful consideration and extensive research before any clinical applications can be contemplated.

2. Neural Interfaces and Brain-Computer Interaction (BCI):

- **Understanding the Brain:** We begin by exploring the basic functioning of the brain and its complex network of neurons, the fundamental units of communication. This section will provide a foundation for understanding how BCIs interact with the brain and translate neural activity into meaningful signals.
- **Types of Neural Interfaces:** We will delve into different types of neural interfaces, categorized by their invasiveness:

Non-invasive BCIs: These interfaces, like EEG (electroencephalography), detect brain activity from the scalp without entering the brain itself. They offer the advantage of being non-invasive and relatively safe, but they also have limited resolution and require significant training for users to control them effectively.

Partially-invasive BCIs: These interfaces, like ECoG (electrocorticography), are placed on or near the surface of the brain, providing higher resolution data compared to non-invasive methods. This allows for more precise control and a wider range of applications, but it also comes with increased risks of infection and requires surgery for implantation.

Fully-invasive BCIs: These interfaces, like deep brain stimulation (DBS), are implanted directly within the brain, offering the highest level of resolution and control. They have the potential to revolutionize various fields, from restoring lost motor function to treating neurological disorders. However, they also carry the highest risks of infection, bleeding, and damage to brain tissue, necessitating careful evaluation of the potential benefits and risks before implantation.

- **Decoding and Encoding Brain Signals:** This section explores how BCIs translate brain signals into commands for external devices or vice versa. We will discuss the significant advancements in decoding and encoding techniques, allowing for more intuitive and natural interaction with technology. Decoding involves interpreting the complex patterns of neural activity into meaningful commands, while encoding involves translating external signals into a form that the brain can understand and utilize. As these techniques continue to evolve, the potential applications of BCIs in human augmentation become increasingly diverse and sophisticated.

3. Artificial Intelligence and Machine Learning Integration:

- **The Role of AI in Augmentation:** This section explores how AI and machine learning play a crucial role in developing and optimizing augmentation technologies. AI algorithms can analyze vast amounts of data from brain scans, user interactions, and other sources to:

Personalize interventions: By tailoring augmentation strategies to individual needs and preferences, AI can ensure that interventions are more effective and have a greater impact on user experience.

Optimize device performance: AI can continuously analyze user data and device performance to identify areas for improvement, leading to more efficient and user-friendly augmentation technologies.

Develop intelligent prosthetics: By incorporating machine learning algorithms, prosthetics can learn and adapt to user behavior, movements, and intentions, providing a more natural and intuitive control experience.

- **Augmenting the Mind with AI:** This section delves into the potential, yet highly debated, area of directly integrating

CHAPTER 4: TRANSFORMING HEALTHCARE AND MEDICINE

Human augmentation presents a transformative force in healthcare and medicine, offering groundbreaking solutions to address existing challenges and improve overall well-being. The following explores the diverse ways augmentation is impacting the medical landscape.

1. Treating Disabilities and Diseases:

- **Prosthetics and Implants:** Advanced prosthetics and implants are no longer science fiction. Bionic limbs, with their increased dexterity and functionality, are restoring lost mobility and function to individuals with disabilities, empowering them to lead more independent and fulfilling lives. Artificial organs like pacemakers and cochlear implants are bridging the gap between lost functionalities and a healthy life, granting individuals the ability to hear, see, or regulate vital bodily functions.

- **Brain-Computer Interfaces (BCIs):** BCIs offer a glimmer of hope for individuals with paralysis or neurological disorders. By establishing a direct communication channel between the brain and external devices, BCIs allow individuals to control prosthetic limbs, communicate through speech synthesis, or even operate assistive technologies. Imagine someone with locked-in syndrome regaining the ability to communicate and interact with the world through the power of BCIs.

- **Gene Editing and Regenerative Medicine:** Gene editing holds the potential to revolutionize how we treat diseases by correcting the genetic mutations responsible for various conditions. Additionally, advancements in regenerative medicine aim to utilize stem cells and tissue engineering to repair or replace damaged tissues, potentially offering solutions for currently incurable conditions like heart disease, diabetes, and spinal cord injuries. This opens doors to a future where previously untreatable conditions become manageable, improving the lives of millions.

2. Personalized Medicine and Augmentation Therapy:

- **Tailored Augmentation:** With the help of AI and big data analytics, doctors are moving towards a future of personalized augmentation therapies. By analyzing an individual's unique genetic profile and specific needs, healthcare professionals can tailor augmentation interventions for maximum effectiveness. This shift from a one-size-fits-all approach to personalized medicine holds immense promise for improving treatment outcomes and patient experiences.

- **Biomarkers and Predictive Augmentation:** Identifying specific biomarkers associated with disease progression using advanced diagnostics can enable early intervention and preventive measures. For example, monitoring brain activity through BCIs could potentially detect the early stages of neurological disorders, allowing for timely intervention before symptoms manifest. This proactive approach could significantly improve patient outcomes and potentially prevent the progression of debilitating diseases.

- **Augmenting the Immune System:** The future of healthcare might see augmented immune systems utilizing implants or biomaterials to enhance the body's natural defenses against pathogens and diseases. This could lead to improved resistance to infectious diseases and potentially even cancer, empowering the body to fight against illness more effectively.

3. The Future of Human Lifespan:

- **Anti-Aging Therapies and Lifespan Extension:** While still in its early stages, research into anti-aging therapies and lifespan extension explores various approaches, including cellular rejuvenation techniques, gene therapy targeting genes associated with longevity, and even augmented organs and body repair. While ethical considerations and potential risks need careful evaluation, these advancements offer a glimpse into a future where aging might be a treatable condition, extending healthy lifespans and potentially improving overall quality of life.

Ethical Considerations:

The immense potential of augmentation in healthcare comes with crucial ethical considerations that demand careful evaluation:

- **Access and equity:** Ensuring equitable access to these potentially life-changing technologies for all individuals, regardless of socioeconomic status or geographical location, is paramount. No one should be denied access to these advancements due to financial limitations or geographical disparities.

- **Informed consent and patient autonomy:** Individuals should have the right to make informed decisions about augmentation therapies, understanding the potential risks and benefits involved. Transparent communication and comprehensive information are crucial for empowering patients to make decisions that align with their values and personal well-being.
- **Unforeseen consequences:** Careful research and long-term monitoring are crucial to identify

and mitigate potential unforeseen consequences of these emerging technologies. A proactive approach to risk assessment and mitigation is essential to ensure the safety and well-being of individuals undergoing augmentation therapies.

CHAPTER 5: REDEFINING WORK AND THE WORKPLACE

Human augmentation is not just transforming healthcare; it's also making its mark on the world of work, reshaping industries and impacting the relationship between humans and machines.

1. Augmentation in Different Industries:

- **Manufacturing and Construction:** Exoskeletons are providing workers with increased strength and endurance, reducing fatigue and injuries while boosting productivity. Augmented reality (AR) glasses are being utilized for training, maintenance, and remote assistance, streamlining workflows and improving efficiency. This allows workers to access crucial information and instructions directly in their field of vision, minimizing errors and improving task completion times.

- **Logistics and Transportation:** Self-driving vehicles, powered by AI and advanced sensors, are revolutionizing transportation and logistics. Pilots, truck drivers, and other transportation professionals may utilize augmentation technologies to enhance their situational awareness, decision-making abilities, and reaction times in complex situations. For instance, pilots might receive real-time weather data and potential collision warnings directly through augmented displays, improving flight safety and efficiency.
- **Healthcare:** Robotics and AI-powered surgical tools are assisting

surgeons in delicate procedures, improving precision and minimizing human error. Nurses and other healthcare professionals can leverage wearable devices to monitor patient vitals, access medical records, and receive real-time decision support from AI algorithms, leading to improved patient care and faster diagnosis. This allows healthcare professionals to focus on the human aspects of care while leveraging technology to streamline routine tasks and provide data-driven insights.

- **Office Work and Knowledge Work:** Brain-computer interfaces (BCIs) hold the potential to revolutionize communication and information processing in the future. Imagine directly typing documents with your thoughts or accessing information through a mental interface, potentially enhancing productivity and efficiency for knowledge workers. However, careful consideration needs to be given to the potential for manipulation of brain data and the ethical implications of such technology.

2. The Rise of Cyborgs and Human-Machine Collaboration:

The increasing integration of technology into the human body blurs the lines between human and machine, giving rise to the concept of "cyborgs." While full-body cyborgs might still be science fiction, the integration of prosthetic limbs, exoskeletons, and other augmentations is already blurring the lines. This raises important questions about the future of work and human-machine collaboration:

- **Will cyborgs hold certain advantages in the workplace?** Individuals with advanced augmentations might have enhanced physical capabilities, cognitive abilities, or even faster reaction times, potentially creating an uneven playing field in certain professions. This raises concerns about fairness and equity in the workplace, and necessitates the development of regulations or safeguards to prevent discrimination against individuals who choose not to augment themselves.

- **How can we ensure equitable access to these technologies?** If advancements become essential for career success, ensuring everyone has access to them, regardless of socioeconomic background, becomes crucial to prevent widening gaps and potential discrimination. Governments, companies, and educational institutions need to work together to create programs that provide affordable access to training and augmentation technologies, fostering a future where the benefits of human augmentation are distributed equitably.

3. Ethical Considerations and Worker Rights:

As human augmentation evolves, a multitude of ethical considerations and worker rights issues demand attention:

- **Job displacement and retraining:** Automation and AI-powered technologies might replace certain jobs, leading to unemployment and requiring significant retraining efforts for workers whose skills become obsolete. Governments and educational institutions need to invest in robust upskilling and reskilling programs to ensure a smooth transition for individuals whose jobs are impacted by automation, mitigating the negative societal consequences of technological advancement.

- **Data privacy and security:** Wearable devices and BCIs collect vast amounts of data about workers' performance, health, and even brain activity. Ensuring data privacy and security, and establishing clear regulations regarding data ownership and usage, is paramount. Workers need to have control over their data and be informed about how it is being collected, stored, and used. Additionally, robust cybersecurity measures are crucial to prevent unauthorized access to sensitive data.

- **Mental health and the pressure to augment:** Individuals might feel pressured to adopt certain augmentations to remain competitive in the job market, potentially leading to mental health concerns and exacerbating existing inequalities. It's crucial to address the potential psychological impact of these technologies and promote a culture of acceptance and inclusivity in the workplace, ensuring that individuals are not pressured to augment themselves if it goes against their personal values or financial limitations.

CHAPTER 6: RETHINKING HUMAN POTENTIAL AND IDENTITY

As we delve deeper into human augmentation, try to understand its profound impact on the very core of our existence: our understanding of human potential and identity.

1. The Blurring Lines between Human and Machine:

The integration of technology into the human body raises profound questions about the boundaries between human and machine. With prosthetics becoming increasingly sophisticated and BCIs offering the potential for direct brain-computer interaction, the lines separating the two become increasingly blurred. This raises several questions:

- **What defines a human?** If our bodies and minds become increasingly intertwined with technology, where do we draw the line between human and machine? Traditional definitions based solely on biological characteristics might need to be re-evaluated as technology becomes an even more integrated part of the human experience.

- **The ethics of human enhancement:** As augmentation allows us to push the boundaries of human capabilities, ethical considerations become paramount. Should we enhance physical strength, cognitive abilities, or even lifespan? If so, who gets to decide, and who has access to these advancements? Careful consideration is needed to ensure equitable access, prevent discrimination, and avoid unforeseen consequences of altering the fundamental nature of what it means to be human.

- **The psychological impact of augmentation:** How will individuals grappling with their identities in the face of advanced augmentation technologies? Will some feel pressure to augment themselves or risk being left behind? Will there be a sense of alienation or anxiety associated with not being "enhanced"? Addressing the psychological aspects of human augmentation is crucial to ensure individuals' well-being and foster a sense of acceptance and inclusivity in an increasingly augmented world.

2. The Impact on Individual and Societal Values:

Augmentation holds the potential to reshape not only individuals but also societal values and norms. Here are some key considerations:

- **Changes in social interaction:** How will human-machine interaction evolve as technology becomes more integrated with our daily lives? Will face-to-face communication and social interaction diminish as we rely more on technology-mediated connections? It's crucial to ensure that advancements in technology do not come at the cost of genuine human connection and robust social interactions.
- **Shifting definitions of success and worth:** In a world where augmentation can enhance physical and cognitive abilities, how will we define success and self-worth? Will

individuals be judged based on their natural capabilities or the level of augmentation they have adopted? Fostering a society that values diversity and individual contributions, regardless of one's level of augmentation, will be critical to maintaining a sense of fairness and inclusivity.

- **Potential for social conflict and inequality:** If access to augmentation technologies remains unequal, it could exacerbate existing social and economic divides. Additionally, the potential for misuse of these technologies, particularly in areas of warfare or social control, raises significant concerns. Addressing these issues proactively through dialogue, regulations, and international cooperation is crucial to ensure that augmentation benefits society as a whole and does not exacerbate existing inequalities or pose threats to individual freedom.

3. The Question of Transhumanism:

Transhumanism is a philosophical and cultural movement that seeks to transcend the limitations of the human body and mind through technology. This chapter explores the potential implications of this movement:

- **Extending human lifespan and health:** As discussed earlier, augmentation technologies have the potential to extend human lifespan and improve health. However, this raises questions about resource allocation, overpopulation, and the potential for societal and environmental consequences of a drastically extended human lifespan.

- **Enhancing cognitive abilities:** Beyond lifespan extension, some transhumanists envision the possibility of significantly enhancing human cognitive abilities. This could lead to the emergence of a new class of individuals with vastly superior cognitive capabilities, further widening the gap between the augmented and unaugmented segments of society. Addressing these concerns and mitigating potential risks associated with cognitive enhancement are crucial before venturing down this path.
- **Merging with machines:** Some visions of transhumanism even propose eventual merging of humans with machines, creating a new level of existence beyond our current understanding. While this might be far-fetched at present, it is important to consider the

philosophical and ethical implications of such possibilities to ensure responsible development and guide future advancements in a thoughtful and ethical manner.

CHAPTER 7: ACCESS, EQUITY AND JUSTICE IN AUGMENTATION

While the potential benefits of human augmentation are vast, ensuring equitable access, preventing discrimination, and upholding justice are paramount considerations in this rapidly evolving field. The potential pitfalls of neglecting these considerations are significant, and proactive measures are necessary to navigate the ethical challenges and ensure a future where augmentation benefits all of humanity.

1. The Risk of Inequality and Discrimination:

Human augmentation, if not carefully managed, could exacerbate existing inequalities and create new challenges:

- **Socioeconomic disparity:** The cost of augmentation technologies is likely to be high initially, potentially creating a situation where only the wealthy have access to these advancements. This could widen the gap between the rich and the poor, creating a two-tiered society where the augmented hold a significant advantage over the unaugmented in areas like employment, education, and healthcare. This could lead to societal unrest and hinder innovation as the pool of talent and diverse perspectives shrinks.
- **Discrimination:** Individuals who choose not to augment themselves, or who cannot afford to do so, might face discrimination in various aspects of their lives. This could include employment discrimination, where employers favor augmented candidates based on perceived

advantages in productivity or skill, or even social discrimination, where individuals face prejudice or exclusion based on their augmentation status. To mitigate these risks, it is crucial to establish laws and regulations that protect individuals against discrimination based on their augmentation status, fostering a culture of inclusivity and respect for diverse choices.

- **Access to education and skills:** As the workplace evolves alongside augmentation, access to education and skills training will be critical to ensure that everyone can participate in the new economy. Governments and educational institutions need to work together to create programs that equip individuals with the skills necessary to thrive in an augmented future, regardless of their socioeconomic background. This could involve reskilling and upskilling initiatives to bridge the gap between existing skill sets and the demands of the evolving job market, ensuring that everyone has the opportunity to contribute and benefit from technological advancements.

2. Ensuring Accessibility and Affordability:

To mitigate the risks of inequality and ensure that everyone benefits from the potential of augmentation, several strategies need to be implemented:

- **Government funding and subsidies:** Governments can play a crucial role in making augmentation technologies more affordable by providing targeted subsidies or establishing insurance coverage for specific augmentation procedures.

This could involve creating financial assistance programs for low-income individuals or families, or exploring public-private partnerships to develop sustainable financing models for these technologies. Additionally, investing in research and development, particularly in open-source initiatives, can accelerate innovation and bring down the cost of these technologies, making them more accessible to a wider range of individuals and nations.

- **Open-source development and collaboration:** Encouraging open-source development and fostering international collaboration in research and development can accelerate innovation, reduce costs, and promote knowledge-sharing. This collaborative approach fosters a global knowledge exchange, allowing for faster advancements and potentially leading to more affordable solutions by encouraging competition and preventing monopolies. Open-source development also fosters transparency and accountability, allowing for public scrutiny and ensuring that these technologies are developed with ethical considerations in mind.

- **Ethical investment and responsible business practices:** Investors and businesses need to operate ethically and prioritize the well-being of society as a whole. This includes ensuring fair pricing for augmentation technologies, promoting responsible marketing practices that avoid creating unrealistic expectations or exploiting vulnerabilities, and avoiding predatory lending or discriminatory practices related to augmentation procedures. Upholding ethical standards not only benefits society but also fosters trust and long-term sustainability for businesses operating in the augmentation industry.

3. Global Governance and Regulations:

The development and deployment of augmentation technologies require robust global governance mechanisms and regulations to ensure responsible development and prevent misuse:

- **International treaties and agreements:** Establishing clear international treaties and agreements can help ensure responsible development, prevent the weaponization of augmentation technologies, and promote equitable access at a global scale. These agreements should address concerns about the military use of augmentation, the potential for biosecurity threats, and the ethical implications of cross-border applications of these technologies.

- **Standardization and regulatory frameworks:** Developing standardized safety protocols and regulatory frameworks can ensure the safe and ethical development and deployment of augmentation technologies while fostering innovation and promoting fair competition in the market. These frameworks should establish clear guidelines for research, development, and clinical trials, ensuring the safety and efficacy of augmentation procedures, while also allowing for flexibility and adaptability to accommodate the rapid pace of technological advancements.

- **Public engagement and ethical oversight:** Public participation in discussions about the development and use of augmentation technologies is crucial to ensure that these advancements align with societal values and ethical principles. Establishing independent ethical oversight boards, composed of diverse stakeholders from various sectors, can further safeguard against potential misuse and ensure that these technologies are developed and used in a responsible manner that prioritizes human well-being and societal benefit.

CHAPTER 8: PRIVACY AND SECURITY CONCERNS

As human augmentation integrates technology ever deeper into our lives, concerns about privacy and security come to the forefront. The following explores the potential threats and vulnerabilities associated with augmentation and proposes strategies to mitigate them.

1. Data Protection and Identity in the Age of Augmentation:

- **Data collection and storage:** Augmentations, especially those involving BCIs and wearable devices, generate vast amounts of personal data, including biometric information, brain activity, and even health data. This data raises concerns about:
  - **Privacy violations:** Who owns and controls this data?

How is it stored and secured? Who has access to it? Robust data protection laws and regulations are crucial to ensure individual control over their personal data and prevent unauthorized access or misuse.

- ○ **Identity theft and discrimination:** With such detailed personal data available, individuals become vulnerable to identity theft and potential discrimination based on their augmented status.

- **Data breaches and security vulnerabilities:** Malicious actors might attempt to access or manipulate data collected from augmented individuals, posing risks of:

Identity theft and financial fraud: Stolen personal data could be used for financial fraud or other malicious purposes.

Blackmailing and extortion: Individuals could be blackmailed or extorted based on sensitive information gleaned from their data.

Manipulation and control: In extreme cases, hackers could potentially manipulate or control augmented individuals through their implants or devices, posing significant security risks.

2. Hacking and Manipulation of Augmented Individuals:

- **Vulnerability to cyberattacks:** Augmented individuals, especially those with BCIs, might become vulnerable to hacking attempts. Malicious actors could:

Disrupt or disable implants: This could have severe consequences, especially for individuals relying on these technologies for critical functions like walking or seeing.

Hijack or manipulate brain activity: In extreme scenarios, hackers could potentially manipulate an individual's thoughts or behavior through a compromised BCI, raising serious ethical concerns and posing significant security risks.

Spread misinformation or propaganda: Hackers could exploit vulnerabilities to spread misinformation or propaganda directly into individuals' minds through compromised BCIs, posing a threat to individual autonomy and societal stability.

3. The Potential for Social Control:

The potential for misuse of augmentation technologies also raises concerns about social control:

- **Government surveillance and monitoring:** Governments could use augmentation technologies for mass surveillance, potentially infringing on individual privacy and freedom of thought.
- **Manipulation of public opinion:** Malicious actors could exploit these technologies to spread propaganda or manipulate public opinion, potentially undermining democratic processes and social stability.
- **Exacerbation of existing inequalities:** If access to augmentation remains unequal, it could be used to further control and marginalize certain groups,

exacerbating existing social and economic inequalities.

Strategies for Mitigating Risks:

- **Robust data protection laws:** Implementing strong data protection laws that give individuals control over their data, mandate transparency in data collection and usage, and hold organizations accountable for data security breaches is crucial.
- **Cybersecurity measures:** Implementing robust cybersecurity measures for augmentation technologies is essential to protect against hacking and unauthorized access. This includes regular security audits, encryption of sensitive data, and continuous vulnerability assessments.

- **Public awareness and education:** Raising public awareness about the potential risks and benefits of augmentation technologies is crucial. Individuals need to be informed about their rights, how to protect their data, and how to identify potential security threats.

- **International cooperation:** Addressing the challenges of privacy and security in the age of augmentation requires international cooperation and collaboration. Governments, technology companies, and civil society organizations need to work together to develop global standards and regulations that ensure responsible development and deployment of these technologies.

CHAPTER 9: THE FUTURE OF HUMAN AUGMENTATION: OPPORTUNITIES AND CHALLENGES

Human augmentation stands at a pivotal point, brimming with potential to reshape the human experience yet fraught with challenges that demand careful consideration. The following delves into the long-term vision for human enhancement, emphasizing the critical need for responsible development, societal co-creation, and the preservation of human agency.

1. A Flourishing Future: Beyond Human Limitations

Augmentation, when approached thoughtfully, holds the potential to transcend the limitations of the human body and mind, ushering in a future characterized by:

- **Enhanced health and lifespan:** Imagine a world free from debilitating diseases and extended lifespans, allowing individuals to live longer, healthier, and more fulfilling lives. Advancements in regenerative medicine and personalized medicine, coupled with augmentation technologies, could offer solutions to currently untreatable conditions, improving overall well-being and potentially pushing the boundaries of human longevity.
- **Augmented abilities:** Technological advancements could empower individuals with enhanced physical strength, cognitive function, and sensory perception, opening doors to previously

unimaginable possibilities. Imagine individuals with prosthetics that not only restore mobility but also surpass natural limitations, or individuals with enhanced cognitive abilities tackling complex problems and fostering groundbreaking innovations across various fields.

- **Human-machine collaboration:** Seamless integration between humans and machines could revolutionize various sectors, fostering creativity, innovation, and problem-solving capabilities beyond what either could achieve alone. This collaborative approach could lead to breakthroughs in scientific discovery, advancements in space exploration, and the development of solutions to some of humanity's most pressing challenges.

However, realizing this flourishing future hinges not only on technological breakthroughs but also on a **holistic approach** that considers:

- **Ethical considerations:** Careful and ongoing evaluation of the ethical implications of augmentation is essential to ensure responsible development and prevent unintended consequences. Questions surrounding access, equity, potential discrimination, and the impact on future generations necessitate ongoing discourse and proactive measures to mitigate potential risks.
- **Social impact:** The potential societal impact of augmentation needs to be thoroughly explored, addressing issues like access, equity, and potential social conflicts arising from unequal distribution of these

68

technologies. Proactive measures to ensure equitable access, address potential biases in development, and foster inclusive participation are crucial to building a future where augmentation benefits all of humanity.

- **Human values:** Throughout the development and implementation of augmentation technologies, it's crucial to uphold fundamental human values such as dignity, equality, and autonomy. Respecting individual choices, fostering transparency, and prioritizing human well-being above all else are essential for navigating the ethical complexities of human enhancement.

2. Responsible Development and Societal Co-creation: Charting a Course for the Future

As we embark on this journey of human enhancement, fostering responsible development and societal co-creation is paramount:

- **Public discourse and engagement:** Open and inclusive public discourse is vital to ensure that augmentation technologies are developed and used in accordance with societal values and ethical principles. This necessitates active participation from diverse stakeholders, including scientists, ethicists, policymakers, the general public, and individuals with disabilities who could provide invaluable insights and perspectives.
- **International collaboration:** Addressing the global implications of augmentation requires international cooperation and

collaboration. Sharing knowledge, establishing common standards, and fostering responsible innovation through global partnerships are crucial for ensuring the safe and ethical development of these technologies for the benefit of humanity as a whole. By working together, nations can ensure equitable access, prevent misuse, and collectively shape a future where augmentation serves the greater good.

- **Sustainable and inclusive development:** Sustainable and inclusive development of augmentation technologies is paramount. This includes ensuring equitable access for all individuals, regardless of socioeconomic background or geographical location, and mitigating the potential environmental impact of these technologies. By prioritizing sustainability and inclusivity, we can ensure that the benefits of augmentation are distributed fairly and that the development process does not come at the cost of environmental degradation.

3. The Power of Choice and Human Agency: Preserving the Essence of What Makes Us Human

Throughout the exploration and development of human augmentation, it is crucial to prioritize:

- **Individual autonomy:** Individuals should retain the absolute right to choose whether or not to augment themselves, free from coercion or pressure. This right to bodily autonomy and self-determination is fundamental to upholding human dignity and respecting individual choices. Open and accessible discussions, coupled with comprehensive information and education about the potential risks and benefits of augmentation, are crucial for empowering individuals to make informed decisions that align with their values and personal aspirations.

- **Meaningful human connections:** While augmentation can offer enhanced capabilities, it is vital to safeguard the importance of genuine human connection and interaction. Technology should not replace or diminish the value of empathy, compassion, and social interaction. Fostering a future where human augmentation complements and enriches our social connections, rather than replacing them, is crucial for preserving the essence of what makes us human.

- **Preserving human essence:** As we enhance our capabilities, it is crucial to remember the very core of what it means to be

CONCLUSION:

SHAPING OUR AUGMENTED FUTURE

Human augmentation stands at a crossroads, brimming with potential to transform our lives yet fraught with challenges that demand careful navigation. As we stand at the precipice of this evolving landscape, it is crucial to remember that the future of augmentation is not predetermined but actively shaped by the choices we make today.

This journey of human enhancement necessitates a multi-pronged approach:

1. Responsible Development:

- **Ethical considerations:** Open and transparent discourse surrounding the ethical implications of augmentation is essential to guide development and prevent unintended consequences.

- **Societal impact:** Addressing the potential social effects, including access, equity, and potential biases, is crucial to ensure that augmentation benefits everyone, not just the privileged few.
- **Human values:** Upholding fundamental human values like dignity, equality, and autonomy throughout development and application safeguards that the technology serves humanity.

2. Societal Co-creation:

- **Public discourse and engagement:** Open and inclusive dialogue involving diverse stakeholders is key to ensure technologies align with societal values and ethical principles.

- **International collaboration:** Sharing knowledge, establishing global standards, and fostering responsible innovation through partnerships ensures safe and ethical development for the benefit of all.

- **Sustainable and inclusive development:** Ensuring equitable access, mitigating environmental impact, and fostering inclusivity in development are crucial for a future where everyone benefits.

3. Preserving Human Agency:

- **Individual autonomy:** Respecting the right to choose whether or not to augment oneself, free from coercion, is paramount to upholding human dignity and self-determination.

- **Meaningful human connections:** Safeguarding the importance of genuine human connection and interaction is crucial; technology should enhance, not replace, social interactions.
- **Preserving human essence:** As we enhance our capabilities, we must strive to preserve the core of what defines us: empathy, compassion, and the ability to connect with others on a human level.

The future of human augmentation is not predetermined by technology, but shaped by our choices, values, and collective vision. By prioritizing ethical considerations, fostering societal co-creation, and upholding human agency, we can navigate this complex landscape and ensure that augmentation serves as a tool to empower individuals, enhance our collective well-being, and shape a future that is not only technologically sophisticated but also ethically sound, inclusive, and respectful of human dignity. The path forward is not without its challenges, but by embracing a collaborative, responsible, and human-centered approach, we can guide the future of augmentation towards a brighter future for all.